CAMEL FARMING FOR BEGINNERS

Your Comprehensive Guide to Starting Out Right- Practical Management Techniques:

Successful Farming Strategies

BRIYAN GREENWALT

Table of Contents

CHAPTER ONE..............................9

  What is camel farming? 9

    Why Consider Camel Farming?.......9

    Basic Camel Farming Requirements 10

  Overview of Camel Products (Milk, Meat, and Wool). 10

    Selecting the Right Camel Breeds .11

CHAPTER TWO 13

  Understanding Camels' Behavior and Temperament..............................13

    Health Considerations for Choosing Camels 14

    Preparing the Farm....................15

CHAPTER THREE...........................17

  Shelter and Housing Needs for Camels 17

Watering System and Hydration Needs 17

Feeding and Nutrition Basics 18

Waste Management and Hygiene Practices 18

Equipment for Camel Farming 19

Basics of Camel Handling Equipment 20

CHAPTER FOUR 21

Tools for Grooming and Healthcare ... 21

The value of transportation apparatus. 21

Crucial Infrastructure for Farms 22

Emergency Supplies and Safety Equipment 23

Daily Care Routine: 23

CHAPTER FIVE 25

Feeding Schedules and Dietary Requirements 25

Health Monitoring and Disease Prevention: 26

    Implement parasite management . 26

    Grooming & Hygiene Practices: 26

    Exercise and Social Interaction: ... 27

    Seasonal Considerations (Summer and Winter): 28

    Understanding Camel Reproduction 29

    Reproductive Anatomy and Cycle . 29

    Breeding Methods and Plans. 30

Pregnancy care and the gestation period 31

    Birth and postnatal care 31

    Managing Young Camels (Calves). 32

CHAPTER SIX 33

Harvesting Camel Products..............33

Camel Milk Production.................33

Camel Milk Products and their Benefits 35

CHAPTER SEVEN 37

Wool Harvest and Processing...........37

Slaughter and Meat Processing Guidelines...................................38

Marketing Strategy....................39

Identifying Target Markets...........39

Pricing Strategies for Camel Products: 40

CHAPTER EIGHT 41

Distribution Channels and Sales Outlets
....41

Regulations and Certifications:.....41

Health Issues: 42

Common Diseases and Treatment 43

Vaccine Schedules......................44

CHAPTER NINE..................................45

Parasite Control and Management 45

Nutritional deficiencies....................46

Emergency Care Protocols 46

CHAPTER TEN 49

Frequently Asked Questions (FAQ) ...49

General Questions 49

© 2024 [Briyan Greenwalt]. Reserved all rights.

All content in this book cannot be duplicated, shared, or conveyed in any way, including photocopying, recording, or other electronic or mechanical techniques, without the author's prior consent in writing. The only exceptions are short quotes included in reviews and certain other noncommercial uses allowed by copyright laws.

Disclaimer

The author's study, experience, and understanding of livestock management constitute the basis of the material in this book. Concerning the material contained in this book, the author is neither connected to, nor has any affiliation with, any organization, corporation, or person.

The author makes every effort to ensure that the material is accurate and thorough, but any errors or omissions, as well as any results resulting from the use of this information, are not covered by this statement. We strongly advise readers to consult a specialist for advice unique to their situation.

CHAPTER ONE
What is camel farming?

Camel farming is the process of producing and breeding camels for milk, meat, wool, and, in some cases, transportation. Camels are robust creatures that have adapted to arid and semi-arid environments, making them valuable assets in many parts of the world.

Why Consider Camel Farming?

Camel farming is attractive for a variety of reasons. Camels thrive in severe settings where other cattle suffer, requiring little water and food. Camel farming is economically viable and culturally significant in many locations because of their products, which include nutritious and therapeutic milk, lean meat with minimal cholesterol, and high-quality wool.

Basic Camel Farming Requirements

Camel farming requires appropriate acreage for grazing, housing, and water supplies. Fencing should be strong enough to protect camels from predators. Additionally, understanding camel behavior and health management is critical for maintaining a healthy herd.

Overview of Camel Products (Milk, Meat, and Wool).

Camels provide a variety of valuable items. Camel milk is extremely nutritious, abundant in vitamins and minerals, and used to make a variety of dairy products. Camel meat, which is noted for its low-fat content and distinct flavor, is becoming more popular in worldwide markets. Camel wool, known for its softness and insulation, is used in textiles and handicrafts.

Sustainability and Environmental Benefits:

Camel farming is environmentally viable in dry areas where other agricultural practices may be impracticable. Camels have lower water and feeding requirements than other livestock, making them ideal for locations with limited resources. Their ability to thrive in severe environments helps to protect vulnerable ecosystems and sustain livelihoods in marginal regions.

Selecting the Right Camel Breeds

For those who are new to camel farming, choosing the right breed is essential. Popular options include the dromedary and Bactrian camels, both of which have significant benefits. Dromedaries are well-known for their ability to adapt to arid regions and are widely used to produce milk. Bactrian camels, with their two humps, are ideal for cooler climates and can provide both milk and meat. Climate, available resources, and farming goals should all be considered while selecting a breed. Understanding these qualities allows you to select camels that are best suited to your individual agricultural environment and goals.

CHAPTER TWO
Understanding Camels' Behavior and Temperament

Before beginning camel farming, it is critical to understand the behavior and temperament of these animals. Camels are typically intelligent, robust, and gregarious animals, yet they can exhibit territorial behavior if not managed properly.

Understanding their social dynamics and communication cues, including as vocalizations and body language, is essential for successful handling and management. Beginners who recognize these characteristics can build a pleasant relationship with their camels, resulting in smoother farming operations and better animal welfare.

Health Considerations for Choosing Camels

Ensuring the health of your camels is critical from the beginning of camel husbandry. Before purchasing camels, a veterinarian who specializes in camel care should do a complete health assessment. Vaccinations, parasite control, and overall physical health are all important factors to consider.

Beginning with healthy animals reduces the danger of disease outbreaks and costly interventions later on. Regular health monitoring and preventive care routines should be created to ensure camel well-being throughout their farming career.

Evaluating the Purpose of Camel Farming: Milk, Meat, and Wool

Defining your objectives in camel farming is critical for both strategic and operational success. Whether your goal is to produce camel milk, meat, wool, or a combination of the three, each demands its own management strategies and infrastructure.

For example, dairy businesses require regular milking practices and specific nutrition, whereas meat production necessitates effective development and weight management protocols. Understanding the unique demands and market opportunities connected with each goal allows beginners to adjust their farming methods to achieve long-term and profitable results.

Preparing the Farm

Before you start camel farming, be sure your land fits the essential standards.

Camels require plenty of space to wander and graze, usually at least 1-2 acres per camel. Fencing is critical to securing the perimeter and preventing predator escapes or intrusions. Ensure that the fencing is strong and tall enough to hold adult camels, who can reach heights of up to 7 feet.

CHAPTER THREE

Shelter and Housing Needs for Camels

Providing enough shelter is critical for protecting your camels from adverse weather situations. A basic, durable building with a roof is adequate for providing shade on hot days and shelter from rain or wind. Bedding like straw or sand should be laid on the floor to provide comfort and absorb moisture. Each camel requires approximately 100 square feet of protected room to rest comfortably.

Watering System and Hydration Needs

Camels are extremely effective water savers, but they still require consistent access to clean drinking water. Installing a dependable irrigation system is crucial. Automatic troughs or manual filling stations can be strategically positioned across the farm to ensure that all camels have easy access. Regularly inspect and clean these irrigation stations to preserve cleanliness and prevent contamination.

Feeding and Nutrition Basics

Understanding camel nutrition is essential for sustaining their health and production. Camels are herbivores that mostly eat grasses, grains, and hay. Provide a well-balanced diet high in fiber, supplemented with minerals and vitamins as needed. Monitor their food consumption to ensure they get enough nourishment, especially during breeding or breastfeeding when their nutritional requirements may change.

Waste Management and Hygiene Practices

Proper waste management is critical for providing a clean and healthy environment for your camels. To reduce disease and parasite transmission, remove manure from pens and paddocks regularly. Consider composting manure as a fertilizer for pastures to promote sustainable farming methods. Keep feeding and watering stations clean to reduce contamination and maintain overall cleanliness on your camel farm.

Equipment for Camel Farming

A camel farm requires several crucial pieces of equipment to be managed efficiently. These include food troughs, water troughs, and safe fences to form enclosures. Proper handling equipment, such as halters and leads, is essential for properly moving and restraining camels during everyday duties like grooming and medical procedures. Furthermore, temperature-appropriate shelters are required to protect camels from adverse weather conditions, assuring their safety and comfort.

Basics of Camel Handling Equipment

Camel handling equipment is essential for farm operations, assuring the safety of both camels and humans. Halters are used to lead camels and can be constructed of nylon or leather to ensure durability and comfort. Leads tied to halters give control and guidance, allowing for easier mobility throughout the farm. Grooming items like brushes and hoof picks also help to keep camels clean and healthy, supporting general well-being.

CHAPTER FOUR

Tools for Grooming and Healthcare

Grooming equipment is vital for keeping camels in good condition and look on a farm. Brushes made exclusively for camels assist remove dirt, debris, and loose hair, preventing skin problems and preserving coat quality.

Hoof picks are used to clean and check camel hooves, keeping them free of debris and infections. Regular grooming not only improves the physical condition of the camel but also builds the link between the handler and the animal.

The value of transportation apparatus.
Transportation equipment is essential for safely and efficiently transporting camels between destinations. Trailers or specifically designed transport vehicles with appropriate ventilation and safe

partitioning keep camels happy while traveling. This equipment is essential for transporting camels to veterinary appointments, exhibits, or markets while reducing stress and assuring their safety throughout the journey.

Crucial Infrastructure for Farms

A well-planned farm infrastructure supports camels' everyday operations and welfare.

This comprises large enclosures with solid fences to prevent escapes and keep camels safe from predators. Adequate water and feeding stations are strategically located to ensure simple access to nutrition. Furthermore, shelters such as barns or covered locations protect camels from harsh weather conditions, keeping them healthy and comfortable year-round.

Emergency Supplies and Safety Equipment

On a camel farm, safety equipment, and emergency supplies are essential for properly dealing with unforeseen emergencies.

Handlers should wear protective gear such as gloves and helmets when working with camels. First aid kits filled with vital medical items allow for fast treatment of injuries or illnesses.

Furthermore, emergency response plans specify protocols for dealing with accidents or natural disasters, protecting the safety of both camels and farm workers.

Daily Care Routine: Creating a regular daily care routine is critical to camel health. Begin with a morning feeding to ensure fresh water and proper nutrients. Look for symptoms of disease or pain, such as limping or strange behavior. Clean and replenish water troughs and bedding places as needed.

In the evening, make another check to ensure that all camels are settled and comfortable, and modify food as needed. Cleaning pens and enclosures regularly will help to maintain hygiene and avoid illness spread.

CHAPTER FIVE
Feeding Schedules and Dietary Requirements

Camels require a well-balanced diet to remain healthy and productive. Create a feeding regimen that incorporates high-quality roughage such as hay and fresh pasture.

Supplement with grains or camel-specific feed to suit their nutritional requirements, changing quantities dependent on age, activity level, and breeding status.

Ensure access to mineral blocks containing critical nutrients such as calcium and phosphorus. Monitor intake to avoid overeating or undereating, which can cause stomach problems.

Health Monitoring and Disease Prevention: Constant health monitoring is critical for preventing diseases in camels. Regularly check for signs of illness, such as nasal discharge, coughing, or appetite changes. Maintain immunization programs for common infections such as foot-and-mouth and tetanus.

Implement parasite management techniques such as regular deworming and checking for ticks or other external parasites. Isolate unwell animals right away to avoid spreading infections across the herd.

Grooming & Hygiene Practices: Regular grooming improves camel health and reduces dangers. Brush camels to remove dirt and debris, paying special attention to areas prone to matting, such as the underbelly and tail. Trim hooves to avoid overgrowth and pain, particularly in rainy weather.

To eliminate junk from the ears and eyes, gently clean them with a damp cloth. Keep bedding clean and provide shaded spaces to reduce heat stress during hot weather.

Exercise and Social Interaction: Promoting exercise and social interaction is critical for camel welfare. Allow time each day for free movement in a safe, contained area to improve joint health and muscle tone. Camels are gregarious animals; to avoid loneliness and behavioral concerns, ensuring they engage with other camels frequently.

Rotate pasture areas to provide new grazing options and encourage natural foraging behavior.

Seasonal Considerations (Summer and Winter): Adapting care routines to seasonal fluctuations ensures that camels thrive year-round.

To avoid heat stress in the summer, provide plenty of shade, freshwater, and access to cooling methods such as misters or fans.

Adjust feeding periods to cooler hours of the day and keep an eye out for indications of dehydration or sunburn. In the winter, give shelter from cold winds and rain, using blankets or coats as needed. Increase food to maintain body condition, and adapt foot care to avoid mud-related problems.

Understanding Camel Reproduction

Camel reproduction is a fascinating part of camel farming that is essential for long-term breeding operations.

Female camels, or cows, normally reach sexual maturity at 3-4 years of age, whilst males, or bulls, become sexually active around 4-5 years.

Understanding their reproductive biology entails comprehending the seasonal breeding cycles and physiological changes that govern effective mating and conception.

Reproductive Anatomy and Cycle

Camel husbandry requires a thorough understanding of reproductive anatomy. Female camels have a unique reproductive system that includes a single cervix and uterus. They have an annual breeding cycle that begins in the fall and peaks in the winter, affected by environmental conditions like as daylight and temperature. This cycle has an impact on breeding success because properly timed mating efforts can maximize conception rates.

Breeding Methods and Plans.

Camel breeding requires strategic planning and careful execution. Natural breeding is a frequent procedure in which a mature bull is brought to susceptible cows during their reproductive period. Artificial insemination (AI) can be used for controlled breeding, but it requires knowledge of semen handling and timing insemination to coincide with the female's ovulation period. Both strategies seek to maximize genetic variety and increase herd productivity.

Pregnancy care and the gestation period

Camel pregnancy lasts between 13 and 15 months after successful breeding. Proper nourishment is essential during this time to support both the cow's health and the developing fetus. Regular veterinarian check-ups track the pregnancy's progress and ensure that any difficulties are addressed swiftly. Adequate shelter and protection from harsh weather also help with a smooth gestation.

Birth and postnatal care

Camel birth, also known as calving, usually occurs in the early morning hours following a short labor.

Immediate postnatal care includes making sure the calf is breathing and feeding properly. The umbilical cord is sterilized and treated to avoid infection.

Within hours, the calf begins bonding with its mother, which is critical for imprinting and long-term health. Observing the duo during the first several days confirms that both are adjusting well.

Managing Young Camels (Calves).

Managing young camels entails creating a favorable environment for growth and development. Calves are extremely vulnerable in their early weeks, relying solely on their mothers' milk for sustenance and immune support. Regular checks for symptoms of disease or malnutrition are required.

As they mature, the progressive introduction of solid food and rudimentary training prepares them for integration into the larger herd dynamics, laying the groundwork for their future duties on the farm.

CHAPTER SIX

Harvesting Camel Products.

Camel Milk Production

Camel milk production involves daily milking regimens.

Keep the camel comfortable and secure in a milking location. To ensure hygiene, thoroughly clean the udders and teats before milking. To gather milk, use a disinfected milking bucket or machine. After milking, filter out any contaminants.

To keep camel milk fresh, store it in clean, sealed containers in a cool place or refrigerate it right away. Regular milking ensures a consistent supply of healthy camel milk, which can be consumed or processed into numerous products.

Milking Techniques and Equipment

Camel milking requires cautious treatment to keep them cooperative. Use either manual milking or advanced milking devices built specifically for camels. Position the camel securely, either with its hind legs secured or in a milking chute. To avoid contamination, properly clean the udders and teats before milking.

To prevent the camel from being stressed, milking should be done gently and steadily.

Properly maintained milking equipment, such as stainless steel buckets or milking machines with camel-specific attachments, provides sanitary milk collection and effective milking sessions.

Milk Handling and Storage

To retain the quality of camel milk, it must be handled carefully. After milking, filter it to remove debris and microorganisms. Place camel milk in clean, food-grade containers. To avoid bacterial growth, keep the milk cold between 2°C and 4°C. Check the freshness of stored milk regularly and destroy any ruined batches as soon as possible.

Proper handling and storage procedures guarantee that camel milk preserves its nutritional value and is safe to consume or further process into dairy products.

Camel Milk Products and their Benefits
Camel milk's nutritious benefits and unique composition make it suitable for a wide range of products. Camel milk can be processed into numerous items such as cheese, yogurt, and ice cream.

These products make use of camel milk's high vitamin and mineral content, as well as its lower lactose level when compared to cow milk. Camel milk products are ideal for lactose intolerant people and provide health benefits such as immune system support and better digestion. Experimenting with various recipes and procedures enables the innovative and successful use of camel milk on the market.

CHAPTER SEVEN
Wool Harvest and Processing

Camel wool is harvested by shearing camels once a year during the shedding season. Secure the camel, then use electric or hand shears to gently remove the wool.

Sort the wool by quality and color before properly cleaning it to eliminate dirt and oils. Camel wool can be spun into yarn or used to weave long-lasting fabrics and carpets.

Camel wool is prized for its softness, warmth, and hypoallergenic characteristics, making it an excellent choice for high-end clothes and home furnishings.

Proper harvesting and processing guarantee that camel wool preserves its natural properties and market value.

Slaughter and Meat Processing Guidelines

Camels must be slaughtered for meat by ethical and sanitary guidelines. Choose a certified abattoir or slaughterhouse that can process camels.

Ensure that the camel is handled calmly and humanely during the killing process. Following slaughter, the carcass should be immediately inspected for quality and safety.

Separate camel meat into slices suited for a variety of culinary purposes. Camel meat is lean, soft, and high in protein, making it an excellent choice for both traditional and modern cuisines.

Proper processing and preservation techniques keep camel meat's flavor and nutritional value for consumers.

Marketing Strategy: Effective marketing is essential for selling camel products. Start by identifying your distinctive selling points, such as organic methods or high-quality camel milk.

Use digital channels such as social media and e-commerce websites to reach a larger audience. Engage potential buyers with informative content about camel goods' health and environmental benefits.

Identifying Target Markets: Identifying your target market entails determining who is most likely to buy your camel products. Consider demographics such as age, lifestyle, and health-conscious customers. Research market trends and preferences to better personalize your marketing efforts.

Branding Camel Farm Products: Develop a distinctive brand identity that reflects your camel farm's quality and values. Create a memorable logo and packaging that reflects freshness and authenticity.

Use storytelling to connect with customers emotionally by emphasizing the inherent benefits of camel products.

Pricing Strategies for Camel Products: Determine competitive yet profitable prices for your camel products based on manufacturing costs, market demand, and competition pricing.

Offer several pricing levels based on product variations or container sizes to accommodate varied client budgets.

Regularly assess and alter pricing plans in response to market input and economic trends.

CHAPTER EIGHT

Distribution Channels and Sales Outlets

Choosing the correct distribution channels is Critical for reaching your target market efficiently. Consider selling directly from your farm store, local markets, or collaborating with health food and specialized retailers.

Explore internet sales platforms to broaden your reach beyond your local markets. Ensure compliance with local food safety and labeling standards.

Regulations and Certifications: Understand the regulatory requirements for marketing camel goods, which can vary by area. Obtain the required certifications for food safety, organic labeling, and animal welfare standards. To increase consumer and merchant trust, and ensure compliance with health rules and inspections. Stay updated about regulatory developments so that you may alter your procedures accordingly.

Health Issues:

Camel farming necessitates attentiveness in health management to ensure the well-being of your herd. Common health issues include respiratory infections, foot difficulties, and skin diseases. Regular health exams are critical for early detection and treatment.

For example, antibiotics are commonly used to treat lung infections, but hoof problems may necessitate trimming or infection treatment.

Topical therapies can help manage skin disorders such as mange. Prompt veterinary care is critical to avoiding complications and maintaining a healthy herd.

Common Diseases and Treatment

Understanding prevalent diseases is crucial for successful camel farming.

Foot and mouth disease, brucellosis, and trypanosomiasis all have a substantial impact on herd health.

Treatment procedures vary, but they typically include antibiotics, antiparasitic medicines, and supportive care.

Regular surveillance and prompt diagnosis are crucial for containing outbreaks and limiting economic losses.

Proper quarantine methods for new animals can help keep infections out of your herd.

Vaccine Schedules

Setting up a vaccination regimen is critical for preventing infectious diseases in your camel herd. Vaccines for infections such as foot and mouth disease, brucellosis, and rabies are critical. Consult a veterinarian to develop a vaccination regimen according to local disease prevalence and risk factors. Vaccines should be given at regular intervals to guarantee continued protection. Keeping thorough records of immunizations and booster doses allows you to track the health of individual camels as well as the entire herd.

CHAPTER NINE
Parasite Control and Management

Camels' health and productivity depend on effective parasite management.

Ticks, mites, and intestinal worms are all parasites that can cause serious health problems if not treated.

Implement a complete parasite control program, which includes frequent deworming and the use of acaricides. Rotating pastures and keeping living circumstances clean can help to minimize parasite populations.

Monitoring camel health for indicators of parasite infection, such as weight loss and decreased appetite, enables early intervention and therapy.

Nutritional deficiencies

Camels need a balanced diet to flourish, and dietary deficits can cause health concerns. Common deficiencies include vitamin A, D, E, and mineral imbalances. Feed a diet high in quality forage, supplemented with mineral blocks or feed additives as necessary. Regularly evaluate pasture quality and alter feeding techniques accordingly.

A nutritionist or veterinarian can assist you in creating a feeding plan that is tailored to your camels' individual nutritional requirements. Monitoring camel health and production gives useful information about the efficacy of your nutritional management measures.

Emergency Care Protocols

To properly deal with unexpected health concerns, camel husbandry requires emergency preparedness. Establish emergency care protocols that address injuries, birth difficulties, and acute diseases. Keep a first aid kit packed with necessary items including bandages, antiseptics, and medications. Educate farm workers on how to recognize indications of discomfort and provide basic first aid.

In emergency cases, fast access to veterinary services is critical. Regular exercises and upgrades to emergency plans maintain your camels' health and welfare.

CHAPTER TEN

Frequently Asked Questions (FAQ)

General Questions

What are the beginning costs of establishing a camel farm? Starting a camel farm requires various upfront investments, including fencing, shelter construction, camel purchases, and the acquisition of necessary equipment such as feeding troughs and water systems. Additionally, veterinarian charges and administrative costs for permits and licenses should be taken into consideration.

The actual value varies greatly based on farm size, location, and number of camels, but a simple setup might cost tens of thousands of dollars.

How much space do camels require? Camels require plenty of area to roam and graze. On average, each camel requires 1 to 2 acres of land for grazing, depending on pasture quality and climate conditions. Adequate space is essential not only for their physical well-being but also for decreasing stress and behavioral difficulties within the herd.

What are the primary obstacles of camel farming? Camel farming involves unique obstacles, such as maintaining their distinct dietary requirements, especially in dry places where food may be scarce.

Specialized knowledge is required for health treatment, including prevention of infections such as camelpox and brucellosis. Furthermore, managing and teaching camels, who can be temperamental animals, requires patience and understanding of their behavior.

How long does a camel live? Camels are extraordinarily hardy creatures, having an average longevity of 40 to 50 years. Proper nourishment, healthcare, and a stress-free environment all help them live longer lives.

Female camels live longer than males, with some exceeding 50 years under ideal conditions.

Can camels be trained? Yes, camels may be trained using patient and consistent methods.

Training normally begins when they are young calves, with a focus on fundamental commands such as haltering, guiding, and reacting to stimuli.

Positive reinforcement strategies work best, including prizes like as snacks or verbal praise.

Establishing trust and bonding with the camels is critical for excellent training results.

Training not only helps with handling and management, but it also improves the herd's overall welfare and productivity.

To sum up, "Carmel Farming for Beginners" is a vital resource for anyone wishing to get started in the field of caramel farming.

This extensive guide covers all aspects of Carmel cultivation, from the fundamentals to sophisticated methods for optimizing output and quality.

With a focus on important areas including pest management, irrigation, soil preparation, seed selection, and harvesting, the book makes sure that novice farmers have all the information they need to be successful.

The need to comprehend regional soil types and climate circumstances is highlighted, giving readers the means to customize their farming methods to their particular settings.

The importance of sustainable agricultural practices is also emphasized in the book, which encourages farmers to use techniques that safeguard the environment in addition to increasing crop yield.

The practical approach of "Carmel Farming for Beginners" is one of its best qualities. It simplifies difficult ideas with its comprehensive visuals, step-by-step directions, and real-world situations. Common problems and troubleshooting advice are also included to help novice farmers be ready for any setbacks and ensure they can solve problems quickly and efficiently.

The book emphasizes the advantages of joining local farming networks and taking part in cooperative activities, which further encourage community engagement and knowledge exchange.

This feature promotes a feeling of unity and group development among farmers in Carmel, which helps the farming community as a whole.

In the end, "Carmel Farming for Beginners" is a priceless tool that gives inexperienced farmers confidence and enthusiasm—it's more than just a how-to manual.

The book enables people to start their caramel farming adventure with confidence and excitement by demystifying the process and offering a strong foundation of knowledge, opening the door for a successful and sustainable farming future.

www.ingramcontent.com/pod-product-compliance
Lightning Source LLC
Chambersburg PA
CBHW071844210526
45479CB00001B/281